ISO/IEC 38500

A pocket guide

Second edition

ISO/IEC 38500

A pocket guide

Second edition

ALAN CALDER

IT Governance Publishing

Every possible effort has been made to ensure that the information contained in this book is accurate at the time of going to press, and the publisher and the author cannot accept responsibility for any errors or omissions, however caused. Any opinions expressed in this book are those of the author, not the publisher. Websites identified are for reference only, not endorsement, and any website visits are at the reader's own risk. No responsibility for loss or damage occasioned to any person acting, or refraining from action, as a result of the material in this publication can be accepted by the publisher or the author.

Apart from any fair dealing for the purposes of research or private study, or criticism or review, as permitted under the Copyright, Designs and Patents Act 1988, this publication may only be reproduced, stored or transmitted, in any form, or by any means, with the prior permission in writing of the publisher or, in the case of reprographic reproduction, in accordance with the terms of licences issued by the Copyright Licensing Agency. Enquiries concerning reproduction outside those terms should be sent to the publisher at the following address:

IT Governance Publishing Ltd
Unit 3, Clive Court
Bartholomew's Walk
Cambridgeshire Business Park
Ely, Cambridgeshire
CB7 4EA
United Kingdom
www.itgovernancepublishing.co.uk

The author has asserted the rights of the author under the Copyright, Designs and Patents Act, 1988, to be identified as the author of this work.

First published in the United Kingdom in 2008 by IT Governance Publishing:
ISBN: 978-1-90535-657-7

Second edition published in the United Kingdom in 2019 by IT Governance Publishing:
ISBN 978-1-78778-168-9

ABOUT THE AUTHOR

Alan Calder is a leading author on IT governance and information security issues. He is Group CEO of GRC International Group plc, the AIM-listed company that owns IT Governance Ltd – the one-stop shop for books, tools, training and consultancy on governance, risk management and compliance.

Alan has written extensively on the issues of IT governance, information security and privacy. He is an international authority on ISO 27001, the international security standard, about which he wrote, with colleague Steve Watkins, the definitive compliance guide, *IT Governance – An International Guide to Data Security and ISO27001/ISO27002*. This work is based on his experience of leading the world's first successful implementation of BS 7799 (with the seventh edition published in 2019) and is the basis for the UK Open University's postgraduate course on information security.

Other books written by Alan include *The Case for ISO 27001* and *Nine Steps to Success – An ISO 27001 Implementation Overview* (now in its third edition).[1] He has also helped develop a wide range of information security management training courses that have been accredited by the International Board for IT Governance Qualifications (IBITGQ).

Alan is a frequent media commentator on information security and IT governance issues, and has contributed articles and expert comment to a wide range of trade, national and online news outlets.

He was previously CEO of Wide Learning, an e-learning supplier; of Focus Central London, a training and enterprise council; and of Business Link London City Partners, a

---

[1] For details of these books, see:
www.itgovernance.co.uk/shop/category/itgp-books.

government agency focused on helping growing businesses to develop.

ACKNOWLEDGEMENTS

ISO/IEC 38500:2015 is copyright ISO/IEC 2015. This pocket guide is not a substitute for purchasing and reading the Standard, copies of which can be ordered from national standards bodies or from
www.itgovernance.co.uk/shop/product/iso38500-iso-38500-it-governance-standard.

CONTENTS

INTRODUCTION

IT governance has become a much-discussed topic among IT professionals. It is not well understood by senior managers, company directors, board members and chairmen, however, which is a pity, because IT governance is a key topic for exactly these people.

In *IT Governance: Guidelines for Directors*, I wrote:

> In today's corporate governance environment, where the value and importance of intellectual assets are significant, boards must be seen to extend the core governance principles – setting strategic aims, providing strategic leadership, overseeing and monitoring the performance of executive management and reporting to shareholders on their stewardship of the organisation – to the organisation's intellectual capital, information and IT. A culture of opaqueness is out of line with today's expectation of pro-activity and governance transparency. Boards that treat IT as merely a functional or operational issue simply don't 'get it'; directors who are not pro-active in understanding the strategic importance of, and operational risks in, intellectual capital, information and communications technology, are – at best – a drag on the effectiveness of their boards. As younger companies, controlled and managed by people who have grown up with IT and its possibilities, transform the business landscape, so those boards that fail to respond can expect their businesses to be destroyed – and whether the destruction is piece by piece or wholesale is, in the long run, irrelevant.

ISO/IEC 38500 – the international standard for the corporate governance of information and communication technology – puts boards around the world in a position from which they can take effective action to apply core governance principles to their information and communication technology.

CHAPTER 1: WHAT IS ISO/IEC 38500?

ISO/IEC 38500 is the international standard for the corporate governance of information and communication technology.

There are, broadly speaking, two types of standards[2]:

- A specification that describes requirements that must be achieved (ISO 9001 and the Payment Card Industry Data Security Standard (PCI DSS), for example).
- A code of practice, which is a set of guidelines and supporting information that describe best practice and provide advice on how something might be done (such as ISO 27002 or ITIL®).

A specification sets out clear requirements against which an audit can be carried out. Third-party certification schemes – such as the ISO/IEC 27001 certification scheme – are able to exist because an accredited certification body can carry out an audit against the requirements of the standard to establish whether or not the requirements are being met.

A code of practice, on the other hand, provides guidelines and advice on a given subject, and does not provide a framework against which an audit can be carried out. Organisations that use the standard can deploy any bit (or bits) of it they think appropriate, and in a way that they consider appropriate.

ISO/IEC 38500 is a code of practice that was jointly published by ISO (International Organization for Standardization) and IEC (International Electrotechnical Commission) which, between

---

[2] ISO refers to Type A and B management system standards. Type A standards contain specifications and requirements (such as ISO 9001), and Type B standards provide guidelines and supporting advice (such as ISO 27002). The absence of 'code of practice' in the title of a given ISO standard (as with ISO 38500) does not imply a different classification.

them, form the system for worldwide standardisation. ISO/IEC 38500 was originally prepared by Standards Australia, the Australian national member of ISO, and had the number AS 8015:2005. It was adopted by ISO and IEC under a 'fast track procedure' in 2008 and published to the international community. The Standard was revised and updated in 2015.

ISO/IEC 38500 is a "high level, principles-based advisory standard".[3] It provides "broad guidance on the role of a governing body, [and] it encourages organisations to use appropriate standards to underpin their governance of IT."[4] ISO/IEC 38500 does not, in other words, replace those standards and frameworks (such as COBIT®, ITIL, ISO 27001, etc.) that an organisation may already have deployed for the better governance of its IT; what it does do is provide a coherent framework for ensuring that the board is appropriately involved.

ISO/IEC 38500 is divided into five chapters:

1. Scope
2. Terms and definitions
3. Benefits of Good Governance of IT
4. Principles and Model for Good Governance of IT
5. Guidance for the Governance of IT

It also has a foreword and an introduction, in which the process by which the Standard was created is outlined and the corporate governance context is described.

---

[3] ISO/IEC 38500:2015, Introduction.
[4] Ibid.

CHAPTER 2: THE CORPORATE GOVERNANCE CONTEXT

ISO/IEC 38500 is clear that governance is distinct from management. It identifies the role of an organisation's governing body and aligns that with the governing body's role as described in the G20/OECD Principles of Corporate Governance, as revised in 2004, and in the 1992 Cadbury Report on corporate governance.

"Corporate governance could be thought of as the combined statutory and non-statutory framework within which boards of directors exercise their fiduciary duties to the organisations that appoint them."[5]

The term 'corporate governance' gained prominence in 1978 when used by Robert Tricker in *The Independent Director*.[6] In *Corporate Governance* (1984), he described corporate governance as being "concerned with the way corporate entities are governed, as distinct from the way businesses within those companies are managed. Corporate Governance addresses the issues faced by boards of directors, such as the interaction with top management, and relationships with the owners and others interested in the affairs of the company."

In the Cadbury, Greenbury and Turnbull reports of the 1990s, the UK led the way for the OECD in defining how what is known as directors' duty of care should be exercised.

The Cadbury report's introduction provides a lucid description of the role of corporate governance:

---

[5] Alan Calder, *Corporate Governance: A Practical Guide to the Legal Frameworks and International Codes of Practice*, Kogan Page, 2008.
[6] According to Professor Andrew Chambers, in *Tottel's Corporate Governance Handbook* (2003).

Corporate Governance is the system by which companies are directed and controlled. Boards of directors are responsible for the governance of their companies. The shareholders' role in governance is to appoint the directors and the auditors and to satisfy themselves that an appropriate governance structure is in place. The responsibilities of the board include setting the company's strategic aims, providing the leadership to put them into effect, supervising the management of the business and reporting to shareholders on their stewardship. The board's actions are subject to laws, regulations and the shareholders in general meeting.

The UK Corporate Governance Code (2018) states that the board should "establish a framework of prudent and effective controls, which enable risk to be assessed and managed".[7] This recognises the need for a risk management framework and leaves little room for imprudent risk taking. Directors' duties in the UK are enshrined in statute by the Companies Act 2006.

ISO/IEC 38500 directly addresses the governing body of an organisation, although it does recognise that, in smaller organisations, the members of the governing body may also have roles in management. In this way, the Standard makes itself applicable to organisations of all sizes, regardless of purpose, design or ownership structure.

---

[7] *UK Corporate Governance Code*, 2018, p. 4.

CHAPTER 3: SCOPE, APPLICATION AND OBJECTIVES

This chapter deals with the scope, application and objectives of ISO/IEC 38500. It also sets out some of the benefits of using the Standard in terms of the organisation's conformance and performance.

Scope

As might be expected, the scope of the Standard is "guiding principles for members of governing bodies of organizations […] on the effective, efficient, and acceptable use of information technology (IT) within their organizations".[8] ISO/IEC 38500 recognises that these processes could be controlled by one of the following:

- IT specialists within the organisation.
- External service providers.
- Business units within the organisation.

The Standard is directed at providing 'guiding principles' for members of governing bodies of organisations on how to ensure that the use of information technology within their organisations is effective, efficient and acceptable. It also recognises that it has a role in providing guidance to the wide range of people whose role might be to advise, assist or inform governing bodies – including external specialists and IT auditors.

Application

As is usually the case with standards published by ISO/IEC, ISO/IEC 38500 is written to be sector-agnostic. It is designed so that it can be applied by companies of all sizes and from all sectors: public, private and not-for-profit.

---

[8] ISO/IEC 38500:2015, Clause 1.

Objectives

The Standard aims to "promote effective, efficient, and acceptable use of IT" in three ways[9]:

1. Assuring stakeholders (which includes consumers and shareholders, as well as employees and providers/vendors) that they can have confidence in the organisation's IT governance if the Standard is followed.
2. Informing and guiding governing bodies in their IT governance activities.
3. Establishing a vocabulary for the governance of IT.

Benefits

ISO/IEC 38500 "establishes a model for the governance of IT"[10] and helps governing bodies find an appropriate balance between risk and reward in their stewardship of the organisation's IT investment – exactly the requirement of today's corporate governance regime.

The Standard identifies two principal benefits that organisations can derive from following its guidance.

1. Conformance – directors who exercise proper IT governance are more likely to address specific IT-related risks and compliance requirements (and the Standard provides examples of these) in a way that enables them to demonstrate that their obligations have been met.
2. Directors, though, are not simply responsible for complying with legislation; they also have to take risks and deliver a financial return for their shareholders. In the public and not-for-profit sectors, they must manage the costs of the organisation efficiently to deliver against the expectations of their stakeholders. Directors who apply the guidance of ISO/IEC 38500 are more likely to succeed at this than those who do not. Again, the Standard identifies

---

[9] ISO/IEC 38500:2015, Scope.
[10] ISO/IEC 38500:2015, Clause 3.

a number of ways IT can contribute positively to the performance of the organisation.

Definitions

ISO/IEC 38500 contains a number of definitions of terms used within the Standard. Those dealing with risk are taken from ISO Guide 73:2009. The most important of these definitions provide for the corporate governance of IT, or what most people simply call IT governance:

> The system by which the current and future use of IT is directed and controlled.

The definitions are all good, sensible, practical ones that will make sense to any director or manager and which, on their own, almost justify purchasing a copy of the Standard!

CHAPTER 4: PRINCIPLES AND MODEL FOR GOOD GOVERNANCE OF IT

This, the fourth chapter of ISO/IEC 38500, contains the meat of the matter, the most important part of the Standard, and the core of the Standard's concept of IT governance. It identifies six principles of good IT governance, and three main tasks for which governing bodies are responsible.

Six principles

The six principles – which are intended to guide decision-making – of good IT governance are:

1 Responsibility;
2 Strategy;
3 Acquisition;
4 Performance;
5 Conformance; and
6 Human behaviour.

The principle of **responsibility** recognises that those responsible for IT within organisations must understand and accept their responsibilities in respect of the supply and demand for IT. They must also have the authority to perform the actions for which they are responsible. This principle encompasses the notion of 'accountability'.

Strategy recognises that an organisation's business strategy should take into account current and future IT capabilities; conversely, the IT strategy should reflect the requirements of the business strategy. This notion is often described as business–IT alignment, as though the requirement is a surprising one!

Acquisition is the principle that stakeholders should applaud: it argues that IT investment decision-making should be clear and transparent, with an appropriate balance between cost and opportunity, with a clear understanding of risk and both a long- and a short-term view.

The principle of **performance** recognises that IT should be 'fit for purpose'. IT systems must deliver the planned capacity and capability, and associated risks must be mitigated, to provide the intended benefits. Performance requires ongoing monitoring; IT service management is one way of expressing this principle in action.

IT underpins financial accounting, and houses, supports and manipulates data on which the organisation's survival depends; the principle of **conformance** requires the organisation to ensure that IT complies with all regulatory and contractual requirements; standards such as ISO/IEC 27001 have a key role to play here.

IT, of course, is part of an organisation that depends primarily on its people; the sixth principle, **human behaviour**, requires IT policies, practices and decisions to respect human behaviour (which is one of the defined terms in the Standard).

The IT governance model

ISO/IEC 38500 proposes a model for IT governance, which is set out in Figure 1. This model, which is derived from the original version published in AS 8015:2005, is a clear and simple one that clearly contextualises the governing body's role in respect of IT governance.

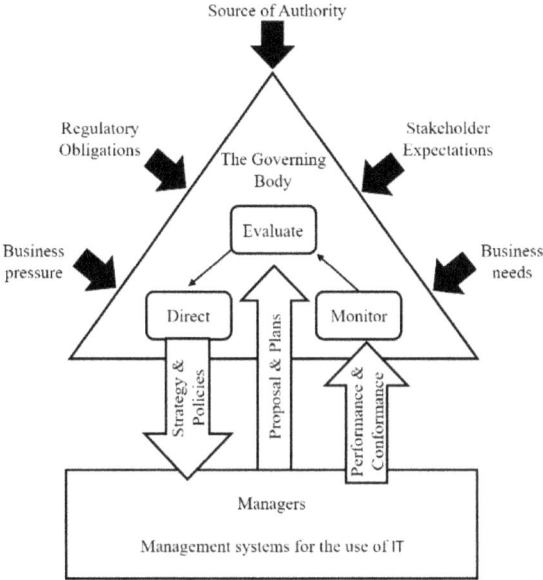

Figure 1: Model for governance of IT[11]

Under this model, governing bodies have three main tasks to ensure effective governance: evaluate, direct and monitor.

Evaluate

ISO/IEC 38500 says governing bodies should evaluate the current and future use of IT (including strategies, implementation plans, supply arrangements, and so on, whether this is internal, external or some combination of both). They should take account of pressures acting on the business, including technological change, economic and other trends, and

---

[11] Original image copyright ISO/IEC 2015.

politics; evaluations should be regular, and be informed by and consider current and future business needs and objectives.

Direct

The governing body must assign responsibility for implementation of IT strategies and policies. It must therefore hold management to account for delivery of those plans. Plans set the direction for IT investment, operation and projects, while policies are directional and should help establish sound behaviour.

This action encompasses the requirement for good, transparent and timely information from management to the board about the progress of IT operations and projects, thus putting the board in a position to ensure that IT projects move smoothly into the operational phase without more disruption than planned for. As IT projects are usually high-risk undertakings with non-trivial consequences in the event of failure or budget excess, this aspect of just this one IT governance action could have a significant effect on improving rates of IT project success.

Monitor

Those directors who want timely information that will enable them to act must first implement monitoring systems that will tell them what is going on – and which will alert them to any failures to comply with regulation, statute or contract. Internal audit is as much a part of effective monitoring as is clear management accountability and meaningful performance reporting.

Accountability

ISO/IEC 38500 makes a very clear statement in this chapter: "accountability for the effective, efficient and acceptable use and delivery of IT by an organisation remains with the governing body and cannot be delegated."[12]

---

[12] ISO/IEC 38500:2015, Clause 4.2.

CHAPTER 5: IMPLEMENTING THE SIX IT GOVERNANCE PRINCIPLES

The fifth chapter of ISO/IEC 38500 describes how the three actions intersect with the six principles; it provides, if you will, guidance on how the six principles are to be implemented, by applying the three actions in each case. Of course, none of this is intended to be exhaustive, and each organisation is encouraged to give "due consideration" to its own nature and make an "appropriate analysis of the risk and opportunities for the use of IT".[13]

Responsibility

Evaluate

- Options for assigning responsibilities.
- The competence of those given operational decision-making responsibilities, with a preference for these to be business managers supported by IT specialists.

Direct

- That strategies are followed according to assigned responsibilities.
- That required information is received.

Monitor

- The establishment of appropriate IT governance mechanisms.
- The acknowledgement and understanding of responsibilities by those that hold them.
- The actual performance of those with responsibilities.

---

[13] ISO/IEC 38500:2015, Clause 5.1.

Strategy

Evaluate

- Developments in IT and business processes to ensure business alignment.
- IT activities to ensure improvements and developments align with the organisation's objectives and satisfy key stakeholder requirements.
- Opportunities to apply best practices.
- To ensure that appropriate risk assessments and risk analysis are carried out (to appropriate international standards).

Direct

- The preparation of strategies and policies that ensure organisational benefit from IT.
- Submission of proposals for innovative use of IT that enable the business to compete and perform better.

Monitor

- The progress of approved IT proposals to ensure they achieve required objectives in required time frames using the resources actually allocated.
- That IT is actually achieving 'its intended benefits'.

Acquisition

Evaluate

- Options for IT to realise business objectives, balancing risk, reward and value for money.

Direct

- That IT assets are appropriately and suitably documented, and with adequate capability to manage the acquired assets.
- Supply arrangements (internal and external) to meet the organisation's business needs.
- The development of a shared understanding of intent between the organisation and its suppliers in any acquisition.

Monitor

- That IT investments produce the required capabilities.
- The extent to which the organisation and suppliers maintain the shared understanding of intent.

Performance

Evaluate

- Management's proposed means for ensuring that IT will support business processes, with the required capability and capacity, taking into account assessed risks and the continuing normal operation of the organisation.
- Risks arising from IT activities.
- Risks to the integrity of the information and protection of information assets, intellectual property and organisational memory.
- Options for assuring effective, timely decisions about the use of IT.
- The effectiveness and performance of the IT governance framework.

Direct

- The allocation of sufficient resources to ensure that IT meets its agreed objectives.
- That correct, up-to-date and secure data is available to support the business.

Monitor

- The extent to which IT actually does support the business.
- The extent to which prioritisation of IT resources actually matches organisational objectives.
- The extent to which IT policies are properly applied and followed.

Conformance

Evaluate

- Regularly the extent to which IT meets the requirements of all applicable regulations, laws, contracts and so on, and conforms with applicable policies and standards.
- The extent to which the organisation conforms to its own IT governance framework.

Direct

- IT management to 'establish mechanisms' and provide regular and routine reports on IT conformance with its obligations.
- To ensure the creation, maintenance and observance of policies and procedures that ensure conformance with those obligations.
- To ensure that staff are professionally developed (e.g. qualifications) and follow guidelines for professional behaviour and development.
- To ensure that all IT actions are ethical (this is about governance, after all).

Monitor

- Internal reporting and IT audit so that reviews are timely, comprehensive, suitable and complete.
- To ensure that all IT activities support the organisation in achieving its full range of obligations, including data protection and privacy, environmental impact, knowledge management and preservation of organisational memory.

Human behaviour

Evaluate

- IT activities to ensure that human behaviours are identified and considered.

Direct

- That IT activities are consistent with human behaviour, which should be obvious but is not always so.
- That there is an effective IT whistle-blowing regime in place, such that risks or concerns from anywhere in the organisation can be drawn to the governing body's attention.

Monitor

- That appropriate attention is given to human behaviour.
- That work practices are "consistent with the appropriate use of IT".

CHAPTER 6: ISO/IEC 38500 AND THE IT STEERING COMMITTEE

ISO/IEC 38500 is a principles-based standard. It describes what governing bodies should do, but does not provide guidance on how they should go about implementing an IT governance framework.

The governing body, in effect, needs to create a mechanism through which it can exercise its IT governance responsibilities and provide the business with technology leadership. The most effective way of doing this is through the creation of a standing board IT committee. Technology or IT leadership requires a specific mechanism of this sort, in a way that, for instance, neither HR nor sales do, for two reasons:

1. HR, sales, marketing, and so on are usually already dealt with effectively as part of the existing governance agenda; most members of the governing body understand the issues around sales and marketing, and the people involved in making sales happen get a great deal of informed attention. The organisation almost certainly already has well-developed governance frameworks for these key activities. No extra benefits would accrue to the organisation through the creation of additional leadership mechanisms for these activities.

2. IT, in contrast, is not as well understood at this level and there are usually no established IT governance frameworks inside organisations. It is not well understood, but it is critical: in 2019, the median investment in IT accounted for 19% of an organisation's annual capital investment, with 49% of organisations increasing their budgets (and 29% reducing them).[14] There is, in other

---

[14] *IT Spending & Staffing Benchmarks 2019/2020*, Computer Economics, *www.computereconomics.com/page.cfm?name=it-spending-and-staffing-study*.

words, a gap between the importance of IT and the understanding of IT: an IT governance framework closes that gap, providing all those with a limited understanding of IT in the enterprise with a framework within which they can improve their understanding to a level appropriate for this critical contributor to their competitive position.

The top-level IT steering or strategy committee has a number of functions, some of which (depending on the size, structure and complexity of the organisation) may be dealt with through subcommittees.

This committee takes the lead, on behalf of the governing body, in dealing with IT governance principles (including the decision-making hierarchy), strategy and risk treatment criteria. ISO/IEC 38500 is very clear in its statement that the governing body cannot escape its overall responsibility for IT and, therefore, it continues to have a key role in monitoring and oversight across the whole of IT, and particularly in respect of project governance.

This monitoring component means that the IT committee has similarities to the audit committee and, given the extent to which IT governance issues impinge on audit issues (particularly around internal control), there is some sense in having a number of members of each committee in common.

They are not necessarily the same committees, however. Many governing bodies expect their audit committees to carry out, on their behalf, the crucial monitoring activities of their overall governance framework. In many such organisations, the monitoring component of the IT governance framework will be included in the agenda of the audit committee to ensure a clear segregation between those responsible for determining (the 'direct' and 'evaluate' actions) the ICT strategy of the organisation and approving investment, and those responsible for monitoring and overseeing the appropriateness and effectiveness of those decisions.

Composition of the IT steering committee

The composition of the IT steering committee should be straightforward. The chair should be selected on exactly the same basis, following the same rules, as the chair of the audit committee. There should be a majority of independent directors on the committee, and key executives should be invited to attend: the CEO, the CFO and the CIO (or equivalent) would be included as a minimum. In some organisations, it would be appropriate to include the CCO (chief compliance officer) as well.

The other key business heads (whether they are from production, procurement, retail, sales, marketing, and so on depends on the sector, the organisation and the existing management structure) – the ones who would be included in any business strategy committee – should be included in the IT steering committee.

The CIO's position and level of accountability should be clear. The CIO should be on the same level, and have the same status, as the CFO and the other functional heads (e.g. sales, marketing, etc.), with direct responsibility for managing the IT operations and personal accountability for the success of organisational IT activity.

- The IT steering committee needs at least one independent director who has the right mix of business and IT experience, and sufficient gravitas to lead the board's IT governance efforts.
- All the other non-executive directors should be prepared and determined to question (evaluate and monitor) every aspect of IT planning and activity.
- The executive – particularly the CIO and the IT management – should be banned from using IT jargon, and forced to express everything they have to say about IT in a format that focuses on comprehensible (to the non-IT specialist) opportunities, issues, risks or plans.
- The IT steering committee should have access to external, professional advice on this as on other matters. Employ outside experts (strategic IT consultants) as board advisers with the specific brief of confirming that what the board

has been told is accurate, complete and true and, if not, what has been left out.

CHAPTER 7: PROJECT GOVERNANCE

It may seem unusual, in a pocket guide on the international IT governance standard, to have a chapter on project governance. Effective project governance, though, is one of the areas in which ISO/IEC 38500 can have the most immediately beneficial impact, and should be a key consideration for the IT steering committee.

Organisations continually upgrade their IT systems or deploy new systems to improve customer service, reduce cost, improve product or service quality, and deliver new products, services and business models. These deployments often involve strategic risk for the organisation; they always involve operational risk.

Risk management is a responsibility of the governing body and, therefore, project governance – from inception through to deployment – must also be its responsibility; the 'evaluate' action under both the strategy and acquisition principles of the Standard is for directors to subject IT and IT proposals to appropriate risk assessment.

Increasingly, stakeholders are concerned about project failure. In the past, investment analysts were reluctant to assess IT. Institutional shareholders are now becoming more muscular. Technology is as significant a component of the organisation's cost base as its headcount, but usually consumes substantially more capital.

Driven, in part, by the changing corporate governance climate and, in equal part, by the poor record of IT projects, stakeholders and institutional shareholders increasingly seek transparency around IT. This is even more the case in the public sector, where very substantial sums of taxpayer money are at risk in large-scale IT projects. Such projects involve many moving parts and are naturally high-risk activities; while estimates of project failure vary significantly, it remains the case that a non-trivial number of projects either fail entirely, or run over-budget by a considerable margin.

But there is more than project failure involved. A 2004 report estimated that 80% of corporate assets are digital; that number will only have increased in the intervening years, especially considering the rise in the use of personal data and online activity.[15] Indeed, many investors now prefer organisations whose assets are largely or wholly digital, as they are less susceptible to 'traditional' risks, such as depreciation of stocked goods or capital equipment.[16]

Organisations whose IT projects failed usually deployed recognisable project management methodologies; the reasons for failure were invariably to do with failures of project governance rather than simply of operational management.

As shareholders and boards focus on the extent to which information and intellectual capital are fundamental to their competitive position and long-term survival, so they recognise the fiduciary nature of their responsibility to shareholders and other stakeholders in respect of the organisation's information assets and IT.

As they recognise the impact that technology has on business performance (and, consequently, on stakeholder value), so they look increasingly for a framework that ensures that IT projects are aligned with commercial objectives and that enables companies to quantify and report in a consistent manner on IT investments.

IT investment decisions (for *or* against) expose an organisation to significant risk: strategic, financial, operational and competitive. The pace of change is a significant risk. Project risks must be assessed within the organisation's strategic

---

[15] Testimony of Jody R. Westby, PwC Managing Director, to the House of Congress Committee on Government Reform, September 2004.
[16] Barry Libert, Megan Beck and Yoram Wind, "Investors Today Prefer Companies with Fewer Physical Assets", *Harvard Business Review*, 2016, https://hbr.org/2016/09/investors-today-prefer-companies-with-fewer-physical-assets.

planning and risk management framework for the right decision to be made: one that enhances competitive advantage and delivers measurable value. Critically, projects need continual oversight; the assumptions on which they were predicated need continual reassessment and the expected benefits need regular reappraisal.

ISO/IEC 38500 guides governing bodies to "monitor the progress of approved IT proposals to ensure that they are achieving objectives in required timeframes using allocated resources".[17] When an IT project goes off track, governing bodies need to be able to identify this early, and have the understanding and determination to call time on proceedings at the point where it is clear that the project is heading in the wrong direction.

---

[17] ISO/IEC 38500:2015, Clause 5.3.

CHAPTER 8: OTHER IT GOVERNANCE STANDARDS AND FRAMEWORKS

ISO/IEC 38500 is an overarching framework of principles and guidance for the governing body of an organisation. It deals with the governance of IT, not its management.

A number of frameworks and standards have evolved that provide detailed guidance and support for specific areas of IT activity for which the board is responsible. Each of these frameworks has its own strengths and weaknesses, and each is capable of being used on its own, or in conjunction with one or more of the other frameworks; all can be used within an ISO/IEC 38500 IT governance framework.

The most widely recognised frameworks that can help with both conformance and performance include those below.

COBIT – a framework for the governance and management of enterprise IT. At the time of writing, the latest version is COBIT 2019, and there is a wide range of related professional qualifications.

ISO/IEC 27002:2013 – the international code of best practice for information security, and ISO/IEC 27001:2013, the international specification against which an organisation's information security management system can be certified as conforming.[18] There are a range of professional qualifications specifically related to ISO/IEC 27001[19], and widely recognised information security certifications such as CISSP®[20] and CISM®[21] cover much of the same ground.

---

[18] *www.itgovernance.co.uk/iso27001*.
[19] *www.itgovernance.co.uk/shop/category/iso-27001-training-courses*.
[20] *www.itgovernance.co.uk/cissp*.
[21] *www.itgovernance.co.uk/cism*.

ITIL (IT Infrastructure Library®) – an integrated set of best-practice recommendations for IT service management. While ITIL 4 was released in 2019, the earlier version is still very much in use around the world.[22] There is a well-structured and comprehensive framework of professional certifications for ITIL, which has hundreds of thousands of registered practitioners worldwide.

ISO/IEC 20000 – the associated certification standard for IT service management, and heavily based on ITIL. Professional certifications are available.

Business continuity management is an essential component of IT governance, just as it is of corporate governance generally. ISO 22301[23] is currently the world's only formal standard for business continuity management. It provides both a specification and a code of practice that can be effectively used within the context of an ISO/IEC 38500 IT governance framework.

Project management expertise has two main strands: the PMBOK® (Project Management Body of Knowledge) promoted by the Project Management Institute,[24] and the PRINCE2® (Projects in Controlled Environments) school,[25] which was begun by the UK Office of Government Commerce and now incorporates MSP (Managing Successful Programmes)[26] and MoR (Management of Risk),[27] which, between them, provide a solid discipline for the effective management of IT projects. Both project management schools are supported by a structured range of professional qualifications.

Enterprise IT architecture is a key part of effective IT governance and is a specialist discipline that directors may choose to consider early on. The two that are most valuable are

---

[22] *www.itgovernance.co.uk/itil.*
[23] *www.itgovernance.co.uk/iso22301-business-continuity-standard.*
[24] *www.itgovernance.co.uk/pmbok.*
[25] *www.itgovernance.co.uk/prince2.*
[26] *www.itgovernance.co.uk/msp.*
[27] *www.itgovernance.co.uk/M_o_R.*

the Zachman framework[28] and TOGAF® (The Open Group Architecture Framework).[29]

There is a wide range of other specialist standards and frameworks for IT management, dealing with issues ranging from capability maturity models and quality management through to procurement and operations frameworks. See below[30] for a comprehensive list of frameworks and associated information.

Conformance

Principle 5 of ISO/IEC 38500 states that directors should ensure that their use of IT meets all the requirements of applicable regulations and laws, as well as contractual obligations. The mass of regulation (data protection, privacy, anti-spam, internal control, computer misuse, etc.) relating to organisations is complex and ever-changing. While a number of the standards and frameworks described above will help, it is important to identify the specific regulatory requirements of all those laws and regulations that might apply to the organisation, and to ensure that appropriate conformance actions are taken.

As the regulatory environment becomes more complex, it is increasingly sensible to consider methods that cover multiple requirements, where possible. The IT Governance Cyber Resilience Framework, for example, incorporates cyber security, incident response, risk management and business continuity, and is ideal for organisations subject to wide-reaching regulations such as the General Data Protection Regulation (GDPR) or the Network and Information Systems (NIS) Directive, and all overseen by a comprehensive governance framework.

---

[28] *www.zifa.com/*.
[29] *www.opengroup.org/togaf*.
[30] *www.itgovernance.co.uk/frameworks*.

CHAPTER 9: INTEGRATING FRAMEWORKS

Over the years, the word 'holistic' has become associated with a wide range of causes and concepts, often ecological ones, and more recently it has begun to filter through into the business world, as organisations are encouraged to take a more rounded view of their place in the world and their contribution to it.

Yet, for all the opprobrium it receives – occasionally justified – the notion that all aspects of the organisation are interconnected, and that the success (or failure) of one aspect in turn influences others, is surely common sense. All organisations must account for and appropriately integrate their various operating frameworks into a coherent and comprehensible whole if they are to succeed.

Integrating those frameworks is perhaps one of the greatest challenges of effective governance. All too often, frameworks become fiefdoms, operating as silos with little awareness of their impact across the organisation. This can lead to competing and contradictory goals, factional infighting and increased risk to critical projects, as practitioners struggle to demonstrate the supremacy of their particular framework while playing down the achievements of others. Governing bodies are then given inadequate, biased information on which to base their decisions, ultimately subjecting the whole organisation to increased operational risk. Effective integration is the first line of defence against these risks, and should be a primary concern when considering adoption of any new framework or standard.

Of the commonly encountered frameworks, management system standards published by ISO/IEC most lend themselves to effective integration. A common underlying structure gives ISO/IEC management system standards shared ground on which to build an integrated framework, allowing, for example, ISO/IEC 27001 to coexist with and complement other standards such as ISO 9001, ISO 22301, and so on.

These standards share principles and terminology, easing understanding for practitioners. Meanwhile, common management system requirements such as audit and management review allow newly adopted standards to merge effectively with existing systems – all that is required, in many cases, is additional training or a minor increase in headcount to resolve relevant skills gaps. In some cases, third-party audits of ISO/IEC standards can be combined to reduce disruption to operations.

This is not to say that frameworks such as COBIT, ITIL, and the like cannot also be effectively integrated. COBIT, for example, describes an 'evaluate, direct, monitor' approach to governance similar to ISO/IEC 38500. As a detailed governance and control framework, it can be used to implement the principles of ISO/IEC 38500, in conjunction with a continual improvement model to ensure ongoing effectiveness, and supporting ongoing audit and review functions.

Even where frameworks appear to bear little relation to one another, close examination often identifies areas of interaction and dependency that are not apparent at first glance. Such examination can also shed useful light on areas of the organisation that many governing bodies traditionally consider opaque.

The key to all such integration lies in recognising the interdependence of all aspects of the organisation, and accepting that changes in one area will inevitably impact others. Once this concept takes root within the governing body's thinking and begins to cascade down to operational management, integrating apparently disparate frameworks and standards becomes less of a challenge and more of an opportunity.

FURTHER READING

IT Governance Publishing (ITGP) is the world's leading publisher for governance and compliance. Our industry-leading pocket guides, books, training resources and toolkits are written by real-world practitioners and thought leaders. They are used globally by audiences of all levels, from students to C-suite executives.

Our high-quality publications cover all IT governance, risk and compliance frameworks and are available in a range of formats. This ensures our customers can access the information they need in the way they need it.

Our other publications about IT governance include:

- *Governance of Enterprise IT based on COBIT 5 - A Management Guide* by Geoff Harmer, *www.itgovernancepublishing.co.uk/product/governance-of-enterprise-it-based-on-cobit-5*
- *Governance and Internal Controls for Cutting Edge IT* by Karen Worstell, *www.itgovernancepublishing.co.uk/product/governance-and-internal-controls-for-cutting-edge-it*
- *ITIL® 4 Essentials: Your essential guide for the ITIL 4 Foundation exam and beyond* by Claire Agutter, *www.itgovernancepublishing.co.uk/product/itil-4-essentials-your-essential-guide-for-the-itil-4-foundation-exam-and-beyond*

For more information on ITGP and branded publishing services, and to view our full list of publications, please visit *www.itgovernancepublishing.co.uk*.

To receive regular updates from ITGP, including information on new publications in your area(s) of interest, sign up for our newsletter at *www.itgovernancepublishing.co.uk/topic/newsletter*.

Branded publishing

Through our branded publishing service, you can customise ITGP publications with your company's branding. Find out more at
www.itgovernancepublishing.co.uk/topic/branded-publishing-services.

Related services

ITGP is part of GRC International Group, which offers a comprehensive range of complementary products and services to help organisations meet their objectives.

For a full range of IT governance resources, please visit *www.itgovernance.co.uk/it_governance*.

Training services

The IT Governance training programme is built on our extensive practical experience designing and implementing management systems based on ISO standards, best practice and regulations.

Our courses help attendees develop practical skills and comply with contractual and regulatory requirements. They also support career development via recognised qualifications.

Learn more about our training courses and view the full course catalogue at
www.itgovernance.co.uk/training.

Professional services and consultancy

We are a leading global consultancy of IT governance, risk management and compliance solutions. We advise businesses around the world on their most critical issues and present cost-saving and risk-reducing solutions based on international best practice and frameworks.

We offer a wide range of delivery methods to suit all budgets, timescales and preferred project approaches.

Further reading

Find out how our consultancy services can help your organisation at
<u>www.itgovernance.co.uk/consulting</u>.

Industry news

Want to stay up to date with the latest developments and resources in the IT governance and compliance market? Subscribe to our Weekly Round-up newsletter and we will send you mobile-friendly emails with fresh news and features about your preferred areas of interest, as well as unmissable offers and free resources to help you successfully start your project:
<u>www.itgovernance.co.uk/weekly-round-up</u>.

EU for product safety is Stephen Evans, The Mill Enterprise Hub, Stagreenan, Drogheda, Co. Louth, A92 CD3D, Ireland. (servicecentre@itgovernance.eu)

www.ingramcontent.com/pod-product-compliance
Lightning Source LLC
Chambersburg PA
CBHW071125210326
41519CB00020B/6422